FRUITS and VEGETABLES

TABLE OF CONTENTS

By Lindsay Snea

A Crabtree Seedlings Book

Crabtree Publishing

crabtreebooks.com

School-to-Home Support for Caregivers and Teachers

This book helps children grow by letting them practice reading. Here are a few guiding questions to help the reader with building his or her comprehension skills. Possible answers appear here in red.

Before Reading:

- What do I think this book is about?
    - *I think this book is about food.*
    - *I think this book is about fruits and vegetables.*
- What do I want to learn about fruits and vegetables?
    - *I want to learn why fruits and vegetables are good for us.*
    - *I want to know more about how many we should eat in a day.*

During Reading:

- I wonder why...
    - *I wonder why red fruits and vegetables are good for our hearts.*
    - *I wonder why most fruit juices are high in sugar.*
- What have I learned so far?
    - *I have learned that fruits and vegetables contain carbohydrates.*
    - *I have learned that we need to eat a variety of colors.*

After Reading:

- What details did I learn about fruits and vegetables?
    - *I have learned that each color is good for different parts of our bodies.*
    - *I have learned that we need at least five servings of fruits and vegetables per day.*
- Read the book again and look for the vocabulary words.
    - *I see the word **carbohydrates** on page 3, and the word **fiber** on page 18. The other glossary words are found on pages 22 and 23.*

FRUITS and VEGETABLES

Fruits and vegetables are an important part of a healthy **diet**. They contain **carbohydrates**.

Carbohydrates give you **energy** for work and play.

It is important that you "eat the rainbow."

This means you need to eat a variety of different colors.

Red fruits and vegetables, such as strawberries, raspberries, tomatoes, and peppers, are good for your heart.

Orange fruits and vegetables, such as oranges, carrots, sweet potatoes, and cantaloupes, are good for your eyes and skin.

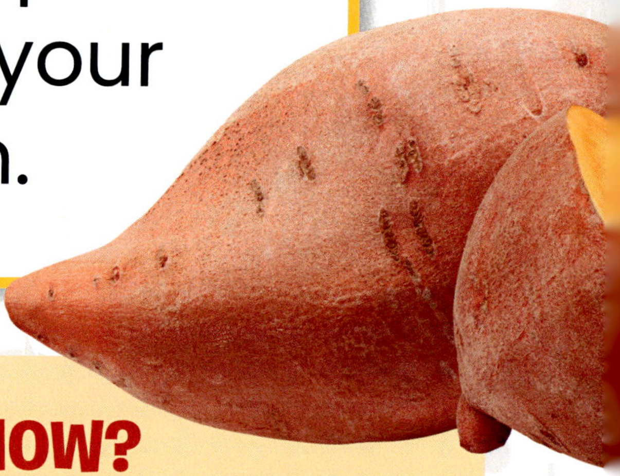

DID YOU KNOW?

Most people think oranges have the highest ***vitamin*** *C content. Fruits and veggies such as kiwis, strawberries, broccoli, and kale actually have more.*

Yellow fruits and vegetables, including lemons, bananas, squash, and mangoes, help keep your **immune system** healthy.

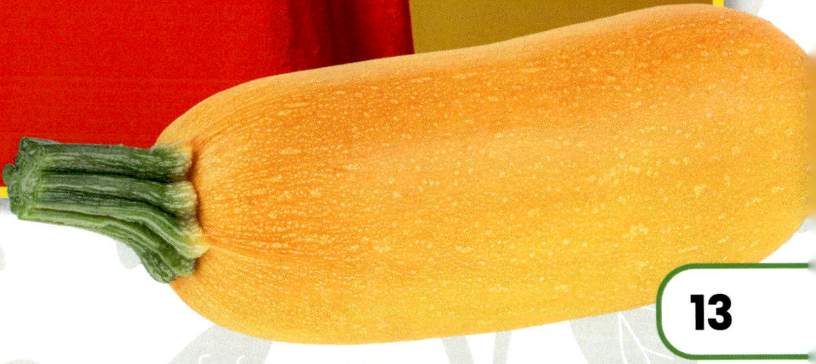

Green fruits and vegetables, such as spinach, cucumbers, kiwis, and green apples, help keep your bones and teeth strong.

Purple and blue vegetables and fruits, such as blueberries, grapes, cabbages, and eggplants, are important for your brain.

It is also important to remember to choose whole fruits over fruit juices. Juices have fewer vitamins and less **fiber** than whole fruits.

DID YOU KNOW?

Most fruit juices have added sugar. Too much sugar in your diet is unhealthy.

You need at least five servings of fruits and vegetables a day. A serving could be one medium apple or two medium carrots.

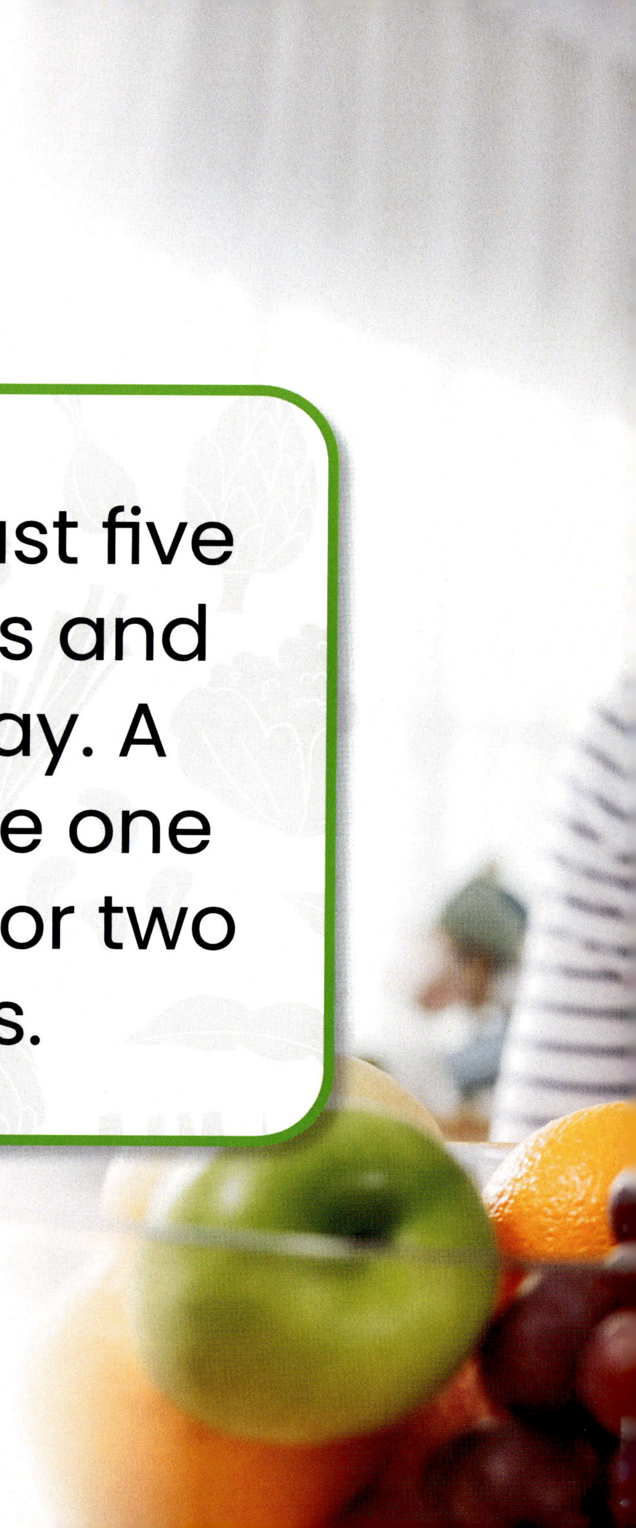

Glossary

carbohydrates (kahr-boh-HAHY-dreyts): Substances found in certain foods that give your body energy

diet (DAHY-it): The food that you usually eat

energy (EN-er-jee): Your body or mind's ability to be active or do work

fiber (FAHY-ber): A material found in fruits and vegetables that helps move food through your body

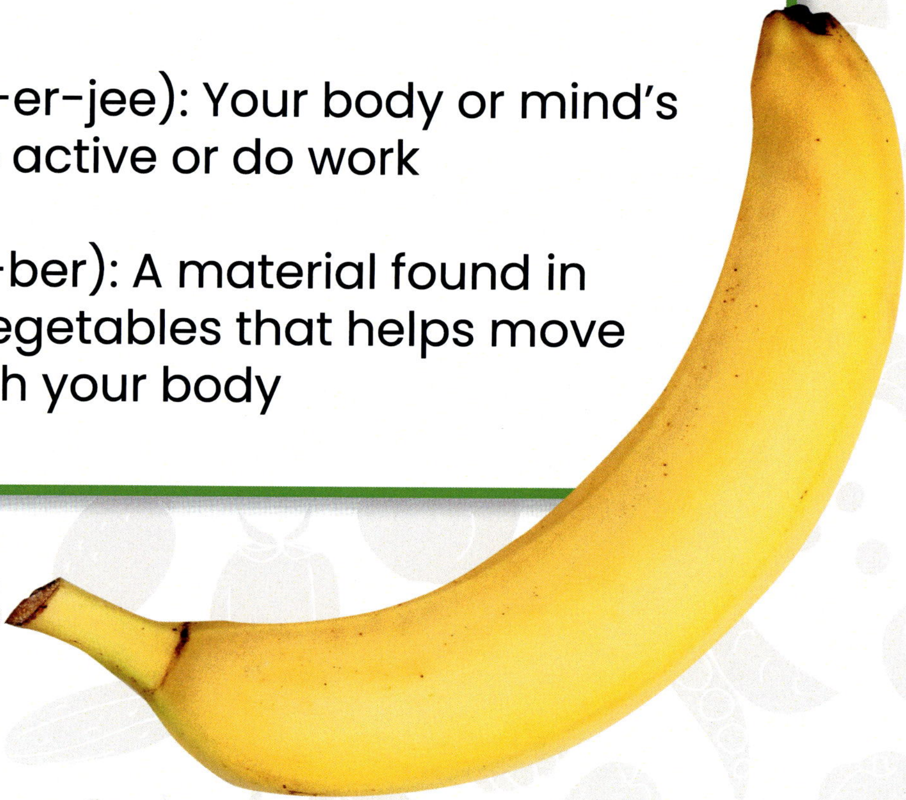

immune system (ih-MYOON SIS-tuhm): The system in your body that protects you from getting sick

vitamins (VAHY-tuh-mins): Nutrients found in foods that help your body grow and stay healthy

Index

About the Author

Lindsay Snea lives in the mountains of Fernie, British Columbia, Canada, with her husband and four children. Lindsay is an elementary school teacher and a lifelong learner who is passionate about cooking, food, fitness, being outdoors, and health.

Written by: Lindsay Snea
Designed by: Margaret Salter
Proofreader: Melissa Boyce
Photographs: All images from Shutterstock
Production manager: Candice Campbell

Crabtree Publishing

crabtreebooks.com 800-387-7650

In Canada: We acknowledge the financial support of the Government of Canada through the Canada Book Fund for our publishing activities.

Printed in the U.S.A./CP112025

Published in Canada
Crabtree Publishing
616 Welland Avenue
St. Catharines, Ontario
L2M 5V6

Published in the United States
Crabtree Publishing
347 Fifth Avenue
Suite 1402-145
New York, New York, 10016

Library and Archives Canada Cataloguing in Publication
Available at Library and Archives Canada

Library of Congress Cataloging-in-Publication Data
Available at the Library of Congress

Hardcover: 978-1-0398-8116-7
Paperback: 978-1-0398-8477-9
Ebook (pdf): 978-1-0398-8236-2
Epub: 978-1-0398-8356-7
Read-Along: 978-1-0398-8561-5
Audio: 978-1-0398-8519-6